**NISTIR 7545**

An Independent Measurement System For Testing Automotive Crash Warning Systems

Sandor Szabo
Joseph Falco
and
Richard Norcross

U.S DEPARTMENT OF COMMERCE
Technology Administration
National Institute of Standards and Technology
Intelligent Systems Division
Gaithersburg, MD 20899-8230

January 2009

U.S. DEPARTMENT OF COMMERCE
Carlos M. Gutierrez, Secretary
NATIONAL INSTITUTE OF STANDARDS AND TECHNOLOGY
Patrick D. Gallagher, Acting Director

Executive Summary

This report describes the National Institute of Standards and Technology's (NIST) role in the Integrated Vehicle-Based Safety Systems (IVBSS) program. IVBSS is a four-year safety initiative sponsored by the United States Department of Transportation (U.S. DOT) to develop and field-test integrated crash warning systems designed to prevent rear-end, lane-change/merge and road departure crashes on light vehicles and heavy commercial trucks.

NIST's primary roles in the program included assisting in the development of verification test procedures, developing a measurement system, and providing field support for vehicle test activities. NIST used the measurement system to collect data for determining whether the prototype warning systems passed or failed all closed-course track tests and for characterizing warning system performance on public roads.

The authors wish to express their thanks to Jack Ference, the National Highway Traffic and Safety Administration IVBSS technical representative and Al Stern of Citizant for their editorial contributions to this report and to Alan Lytle, Kam Saidi and Gerry Cheok of NIST's Building and Fire Research Laboratory for assistance in the static evaluations of the laser scanner.

Disclaimer
Certain commercial entities, equipment, or materials may be identified in this document in order to describe an experimental procedure or concept adequately. Such identification is not intended to imply recommendation or endorsement by the National Institute of Standards and Technology, nor is it intended to imply that the entities, materials, or equipment are necessarily the best available for the purpose.

List of Acronyms

DAS	Data Acquisition System
DVI	Driver-Vehicle Interface
FOT	Field Operational Test
GPS	Global Positioning System
HT	Heavy-truck
IMS	Independent Measurement System
IVBSS	Integrated Vehicle-Based Safety Systems
LCM	Lane-Change/Merge
LDW	Lane Departure Warning
LV	Light-vehicle
MT	Multiple Threats
NHTSA	National Highway Traffic Safety Administration
NIST	National Institute of Standards and Technology
POV	Principal Other Vehicle
RDCW	Road-Departure Crash Warning
RE	Rear-End
SV	Subject Vehicle
U.S. DOT	United States Department of Transportation
UMTRI	University of Michigan Transportation Research Institute

Table of Contents

1 Introduction 4
2 System Overview 4
3 System Components 6
4 Installation 8
5 System Calibration 9
    5.1 Laser scanners 9
    5.2 Road cameras 9
6 Data Collection 10
7 System Measurement Procedures 11
    7.1 Time of warning 12
    7.2 Range and range-rate measurement (R_{FCW} and $Rdot_{SP}$) 12
    7.3 Lateral distance and lateral velocity (LatDist and LatVel) 14
8 Summary of Testing Activities 15
    8.1 Measurement system acceptance testing 16
    8.2 IVBSS warning system testing 18
9 Warning System Performance Evaluation 18
    9.1 R_{FCW} error analysis and correction 18
    9.2 GPS timestamp error analysis 21
10 Summary 21
11 References 23
Appendix A – Forward Warning Range Uncertainty (R_{FCW}) 25
    A.1 Static range uncertainty 25
    A.2 Dynamic range uncertainty measurement 28
    A.3 Time of warning uncertainty 31
    A.4 R_{FCW} combined uncertainty (dynamic and warning time) 31
Appendix B – Lateral Distance Uncertainty (LatDist) 32

1 Introduction

This report describes the National Institute of Standards and Technology's (NIST) participation in Phase I of the Integrated Vehicle-Based Safety Systems (IVBSS) program, a safety research program sponsored by the U.S. Department of Transportation (U.S. DOT). The goal of this initiative is to determine potential safety benefits and user acceptance of integrated rear-end, lane-change/merge and road departure crash warning systems for light vehicles and heavy commercial trucks. Additional information about the program is available on the Internet at the following site:
http://www.its.dot.gov/ivbss/index.htm

NIST's primary roles in the program included assisting in the development of verification test procedures, the design, construction, and characterization of an independent measurement system, and providing field support for vehicle test activities. The verification tests provide an objective means to evaluate warning system performance in a safe and controlled test-track environment. More information on IVBSS performance evaluation and verification tests appear in [1][2][3][4][5]. Additional background on NIST's role in prior DOT crash warning system evaluations appear in [6][7][8].

A critical component of the IVBSS test program is the independent measurement system, which enabled analysis of warning system performance without using warning system components or affecting warning system operation. This document provides details about the NIST independent measurement system (referred to as "the measurement system"), including the design, characterization, and operation during light-vehicle and heavy-truck verification testing. Valuable insights and lessons learned during vehicle testing are also included.

An important step in developing the measurement system was to characterize its accuracy under static and dynamic conditions. Of particular interest were IVBSS test scenarios that required the measurement system to measure range up to 60 m while travelling at 21 m/s (45 mi/h). NIST researchers developed a novel dynamic test to estimate range-sensor error and uncertainty at high closing speeds.

NIST used the measurement system to collect data for determining whether the prototype warning systems passed or failed all closed-course track tests and for characterizing warning system performance on public roads. The measurement system played a vital role in the following IVBSS Phase I testing activities:
- Initial integration and data integrity testing (April 2007)
- Verification testing (September through November 2007)
- Phase I extension testing. (January through March 2008)

2 System Overview

The NIST independent measurement system (IMS) provides objective data for evaluating the performance of crash warning systems. The vehicle-mounted system measures forward and lateral distances to nearby vehicles and objects around it, as well as vehicle position within the lane. It also captures warning system audible output and correlates the warnings with all other measurements. The IMS accomplishes these tasks independently without using the warning system sensors or requiring internal warning system data. In addition, instrumentation of surrounding vehicles, roadside objects or infrastructure is not required. The system includes custom application software to calibrate sensors, to process sensor data, to measure warning-system delays and to evaluate range measurement ccuracy.

Two major requirements influenced the measurement system design; the first was that it must be usable on both closed-course test tracks and public roads. This precluded the use of a GPS-based range measurement approach, since public road tests would require GPS instrumentation on all vehicles, as well as surveyed locations of road infrastructure and all roadside objects. The second requirement was to use the same measurement system for light vehicles and heavy trucks, some with trailers exceeding 15 m in length.

Forward collision warning (FCW) tests require measuring longitudinal range and range-rate to vehicles in the test vehicle's forward path. Lane change/merge (LCM) and road departure warning (RDW) tests require measuring lateral distance and lateral velocity toward lane markings. In addition, LCM tests also require the lateral and longitudinal distance to the vehicle in the adjacent lane. Figure 1 depicts a sample of measurements provided; Table 1 contains definitions of the primary measurement variables used in all vehicle testing.

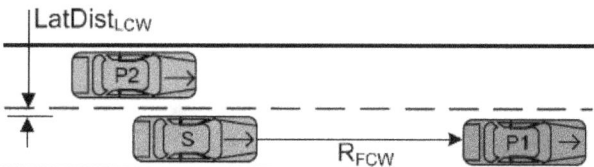

Figure 1. The measurement system uses a laser scanner for the R_{FCW} and calibrated cameras for $LatDist_{LCW}$.

Table 1. Summary of measurement variables.

Name	Unit	Definition
R_{FCW}	m	Range between the subject vehicle (SV) and the principal other vehicle (POV) at the time of a forward collision warning (FCS).
$Rdot_{SP}$	m/s	Range-rate between SV and the POV. (Rdot is commonly used term in the automotive crash warning industry)
$LatDist_{LCW}$	m	Lateral distance between SV front wheel and inside edge of lane boundary at the time of a lane change warning (LCW). Negative values indicate distance outside the lane boundary.
$LatV_{SV}$	m/s	Lateral speed of the SV with respect to lane boundary.

3 System Components

The measurement system uses laser scanners to measure ranges to obstacles in front of and to the side of the test vehicle, and calibrated cameras to measure the distance to lane markers and the vehicle's position within the lane. Figure 2 illustrates the sensor's coverage areas.

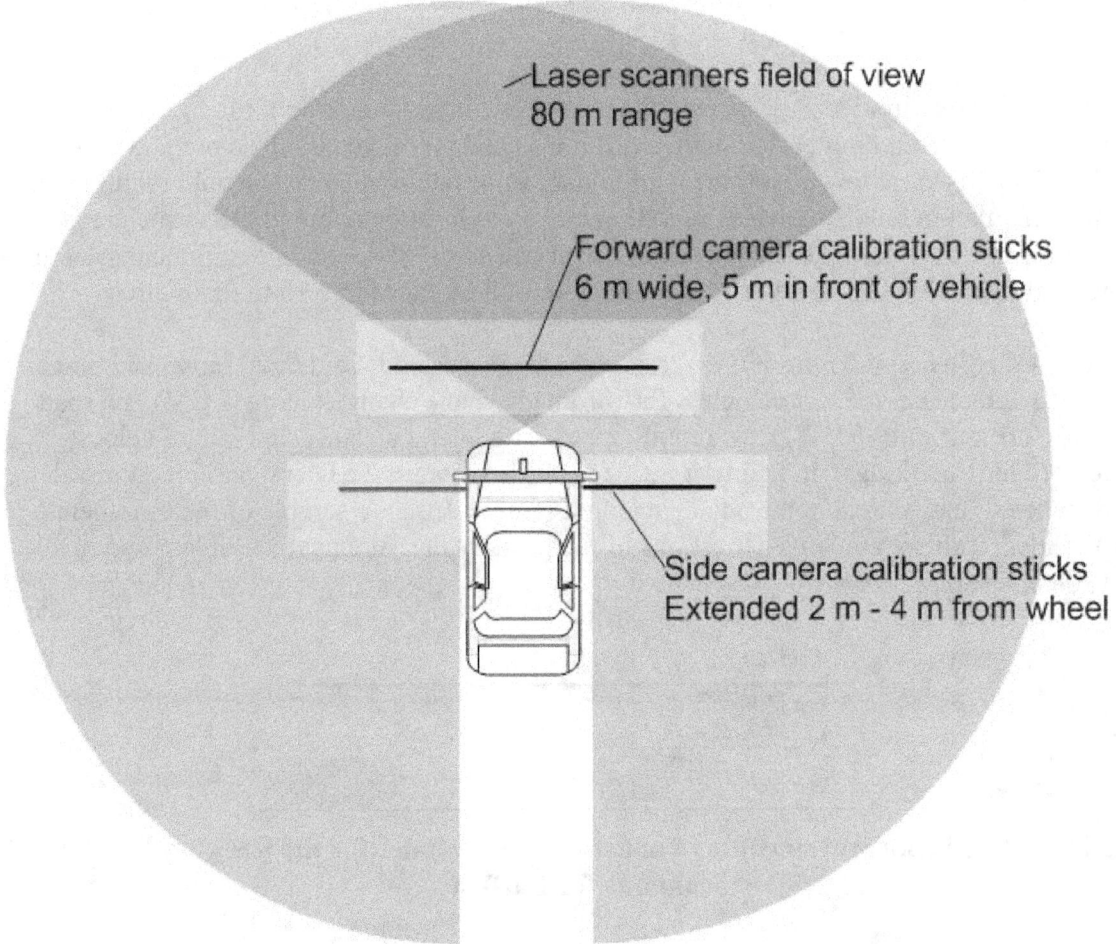

Figure 2. Measurement system fields of view.

Figure 3 shows a system-level diagram of sensors and components and includes those originally developed as part of a prior U.S. DOT–sponsored program,[1] and new components (shaded in green) selected for the IVBSS program. The left portion of the figure shows the external sensors mounted to the front of the car and truck (note that the same sensor package applies to both platforms). The top half of the figure contains the components associated with the laser scanners, while the middle of the figure shows the

---

[1] During the Road Departure Collision Warning (RDCW) Program, NIST developed objective test procedures and an independent measurement system that provided lateral range and lateral velocity to adjacent vehicles, vehicle position within a lane and lateral distance to lane markers [7].

external cameras and GPS time stamping equipment. Data storage devices appear on the right side of the figure. In general, data flows from left to right.

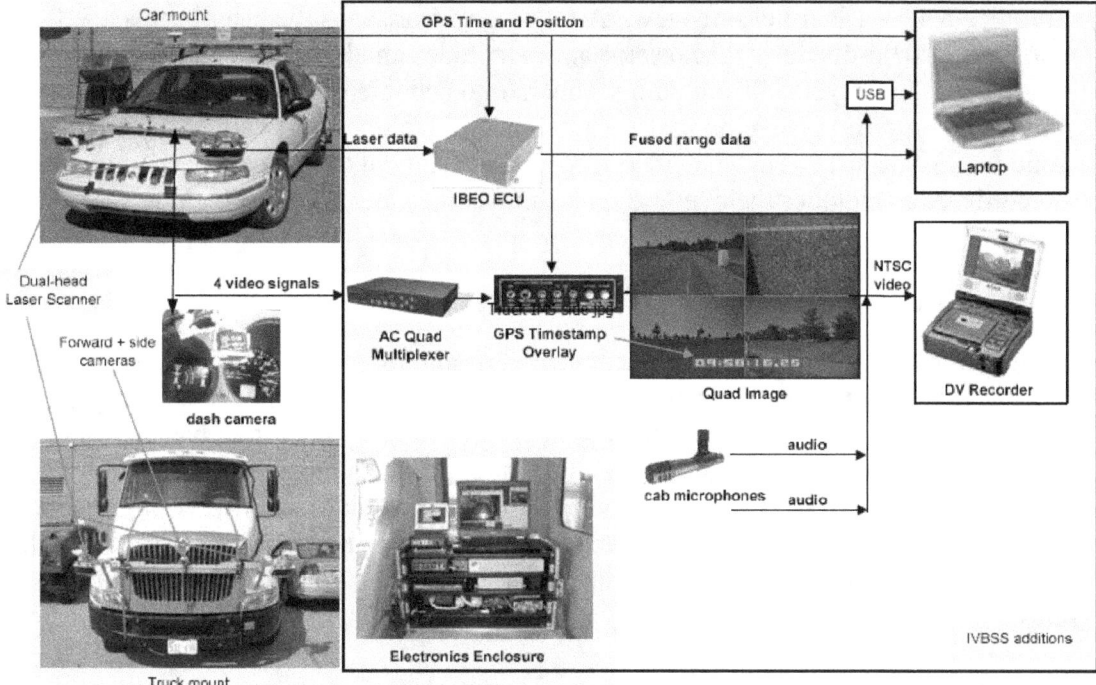

Figure 3. Measurement system components.

The following is a description of the measurement system components and their functions:
1. **Calibrated side- and forward-looking cameras:** There are two side-looking cameras and one forward-looking camera for measuring the distance between the outside edge of a front wheel and the adjacent lane boundary. The forward-looking camera is less accurate, but provides backup measurements when the vehicle is traveling over a lane marker and the marker is not visible in the side cameras. The system includes custom software to calibrate the cameras using specially marked meter-sticks, and to enable users to measure distances to features on the ground by selecting pixels from captured video. Distance measurements are only available along pixels that correspond to the initial position of the calibration sticks.
2. **Cab camera and microphone:** The system uses a camera and microphone mounted in the cab to capture the time of audible warnings and driver actions in the cab.
3. **Camera multiplexer:** A camera multiplexer combines the four camera outputs into a single, quad video-stream. The cameras use 24 v AC power for synchronization.
4. **GPS time-stamp overlay:** A GPS timestamp device overlays local time and frame number (from 0 to 29^2) on each video frame.
5. **DV Recorder**: A digital video recorder captures audio and video at 30 Hz.

---

[2] More precisely, color video frequency is 29.97 frames/s. The timestamp overlay uses a standard drop-frame correction algorithm to maintain the 0-29 frame count.

6. **Dual-head laser scanner**: Commercially developed laser scanners mounted on the front corners of the vehicle provide distances to objects around the test vehicle at ranges from 1 m to 80 m. Each laser scanner uses four fixed lasers mounted to provide a 4° vertical field-of-view. A rotating mirror scans the laser beams a 320° horizontal field-of-view. The laser scanner includes an electronic control unit (ECU) that fuses the range data from both scanners into a single Cartesian coordinate system.
7. **Laptop computer**: A laptop uses commercially developed software to record range and video data at 10 Hz. The video recorded is identical to that captured on the DV recorder (i.e., includes the overlaid GPS time-stamp), but does not include an audio track.

4 Installation

The measurement system's two laser-scanner heads and three video cameras mount to a sensor rack that attaches to the front of the vehicle (see Figure 4). The rack uses pivoting feet to conform to different vehicle body styles and contours. Each foot has three suction cups to eliminate slipping and to provide initial support during installation. On the truck, an additional support strut resting on top of the bumper provides stability for the laser scanners. Straps above and below the rack provide tension and stabilize the sensor rack. The rack also has sliding bars for adjusting the laser-scanners to adapt to different vehicle widths.

External sensors connect to the measurement system electronics via power and signal cables that are bundled and attached to the vehicle body using suction mounts. The electronics are located in the back seat of the vehicle (see Figure 5). The cable assembly, around 2.5 cm in diameter, passes through a gap in the side window, with a foam gasket sealing the gap. The cab video camera and microphone capture audible warnings and driver actions inside the passenger compartment. Installation of sensors and electronics takes approximately 1 hour.

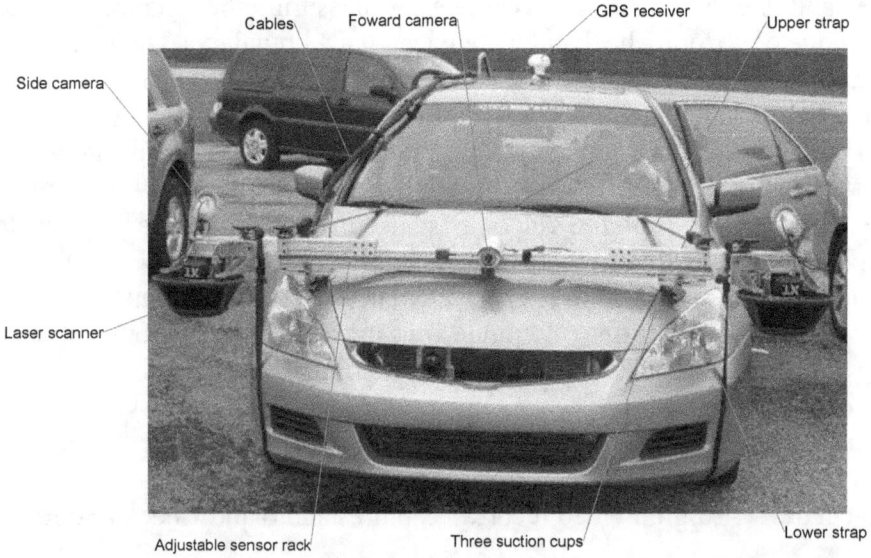

Figure 4. Sensor rack mounted to front of light vehicle.

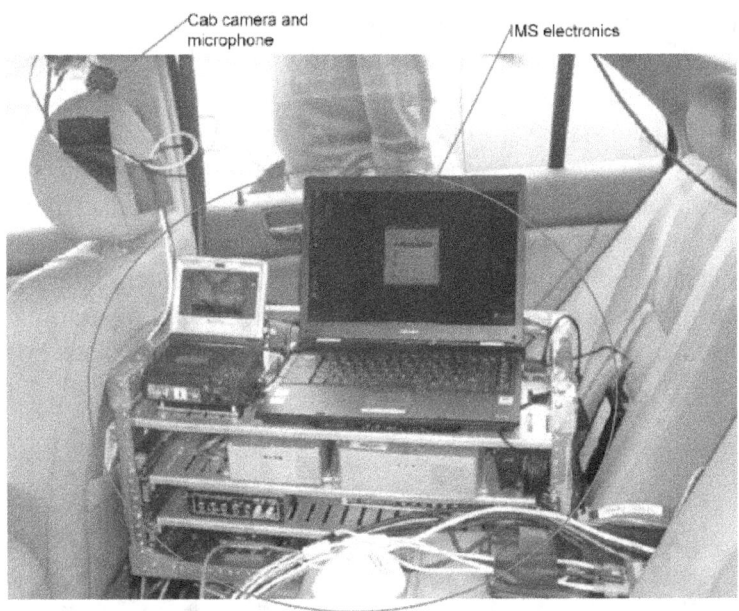

Figure 5. Measurement system electronics in back seat of light vehicle.

5 System Calibration

Physical adjustment and calibration of the measurement system's sensors (laser scanners and external cameras) is necessary to ensure that the sensors cover the required field of view and the data are within the required measurement accuracy. The procedures used provide a good balance between achieving the desired measurement accuracy and minimizing the time required to perform the calibration; typically, calibration requires approximately 1 hour.

5.1 Laser scanners

Once the sensor rack is on the vehicle, the rack's pitch angle and extensions bars are set to ensure the scanners have the correct field-of-view. To fuse data from both scanners into a single coordinate system, the scanner's electronics read a file at start-up that contains calibration parameters defining the rotation and translations between the scanner heads. The rotation value typically remains fixed due to the scanners' fixed mounting bracket. Aligning the lateral distance between the heads initially requires placing a single target within both scanners' fields-of-view and adjusting the translation parameter until the scans align. Afterwards, extension locations of the bar are marked so that subsequent use only requires extending the bar to the previously calibrated location. Finally, a fine-pitch alignment of the entire rack ensures the 4° vertical field-of-view provides the required range data. This consists of placing one vehicle in front and a second vehicle in the adjacent lane to the rear at the maximum distances required during testing, and adjusting the safety straps until the range scan covers both vehicles.

5.2 Road cameras

After completing the laser scanner calibration and tightening the sensor rack straps, the next step is to calibrate the external video cameras. First, two calibration sticks marked in alternating black and white 10 cm segments are placed next to the vehicle's front wheel. Figure 6 shows the view of the left side downward facing camera. The vehicle's wheel appears at the left and the sticks align with the axle and extend laterally from the vehicle. The camera faces rearward in order to see approaching vehicles in the blind spot, which explains why the view appears to be to the right side of the vehicle as one would expect with a camera facing forward. Also shown are small red circles between each segment. During calibration, the user places the circles at the edge of each segment in the image using a custom software application. Later, to obtain lateral measurements, the user clicks on a pixel along the row of circles and the software automatically computes the distance from the wheel using the known 10 cm distance between the red circles.

Figure 6. Calibration of side camera using meter sticks.

6 Data Collection

After calibration, the measurement system is ready for data collection. During tests, commercial software supplied by the laser-scanner manufacturer reads, displays and stores range data and video on a laptop computer in real time. Figure 7 shows the laser scanner data on the left side, and the video from the quad cameras on the right side. The live displays enable users to verify data quality during testing, (e.g., are the scanners aligned properly, have the communications failed, etc.). In addition, the test conductors use the range data to initiate maneuvers during a test. For example, during a lane change test where the subject vehicle changes into a lane with a fast approaching vehicle, the test conductor monitors the range data and instructs the driver to change lanes when the approaching vehicle is within the warning zone.

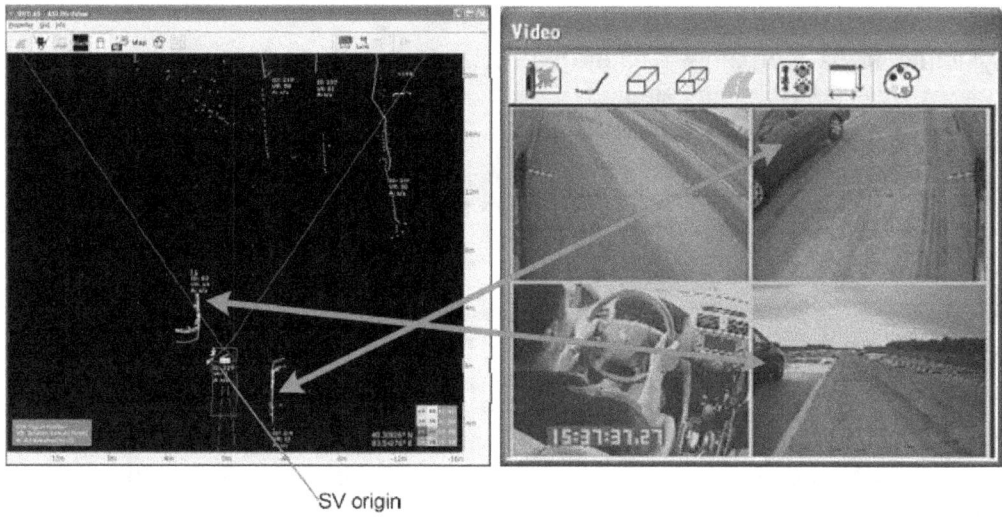

Figure 7. View of laser scanner data collection during a test run.

Besides recording video to the laptop at 10 Hz, a digital video (DV) recorder captures and records video and audio on 90-minute mini-DV tapes at 30 Hz. A commercial package transfers the video from tape into an AVI file in preparation for analysis.

7 System Measurement Procedures

The primary FCW measurement variables in the IVBSS tests are R_{FCW} and $Rdot_{SP}$, the range and range-rate of the subject vehicle (SV) to the principle other vehicle (POV) respectively. The primary LCW and RDW variables are $LatDist_{LCW}$ and $LatV_{SV}$, the lateral distance and lateral velocity of the SV front wheel edge to a lane marker respectively. MT-1, a multiple threat test that requires both forward and lateral measurements, exemplifies the measurement procedures. In the MT-1 test, the subject vehicle (SV) approaches a slower moving vehicle (POV1). At some point, a forward collision warning occurs and the subject vehicle (SV) driver attempts a lane change. However, the subject vehicle (SV) encounters a second vehicle, (POV2), in the adjacent lane resulting in a lane change warning.

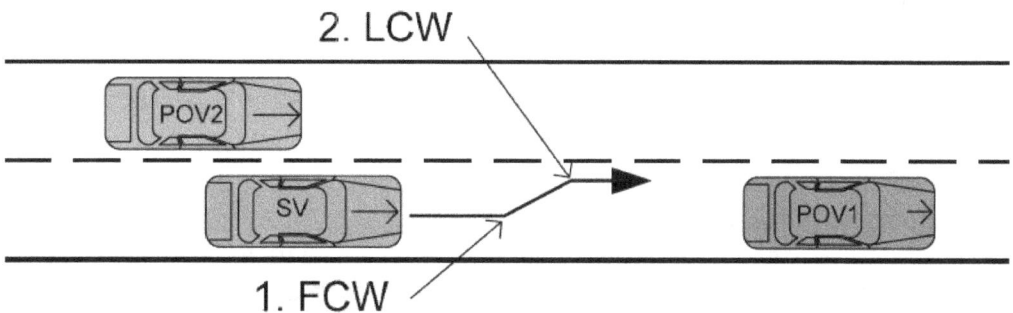

Figure 8. MT-1 test scenario.

The system collects continuous data during each test run. Post-processing and analyzing the data involves the following steps:
- Determine the forward collision and lane change warning times using the captured video and audio
- Measure R_{FCW} and $Rdot_{SP}$ using the laser scanner data
- Measure $LatDist_{LCW}$ and $LatV_{SV}$ using the calibrated side-camera video data

7.1 Time of warning

Figure 9 shows use of a commercial video editing package to locate the time of a warning during an MT-1 test scenario. The audio trace at the bottom of the screen shows both the forward and lane change warnings. The user steps through each frame until hearing the start of a warning. In this multiple threat test example, the forward collision warning starts at frame 15:37:34:13 and the lane departure warning occurs 1.833 s later at 15:37:36:08. This is the actual time the driver first hears a warning in the cab to within 33 ms ($1/30^{th}$ of a second).

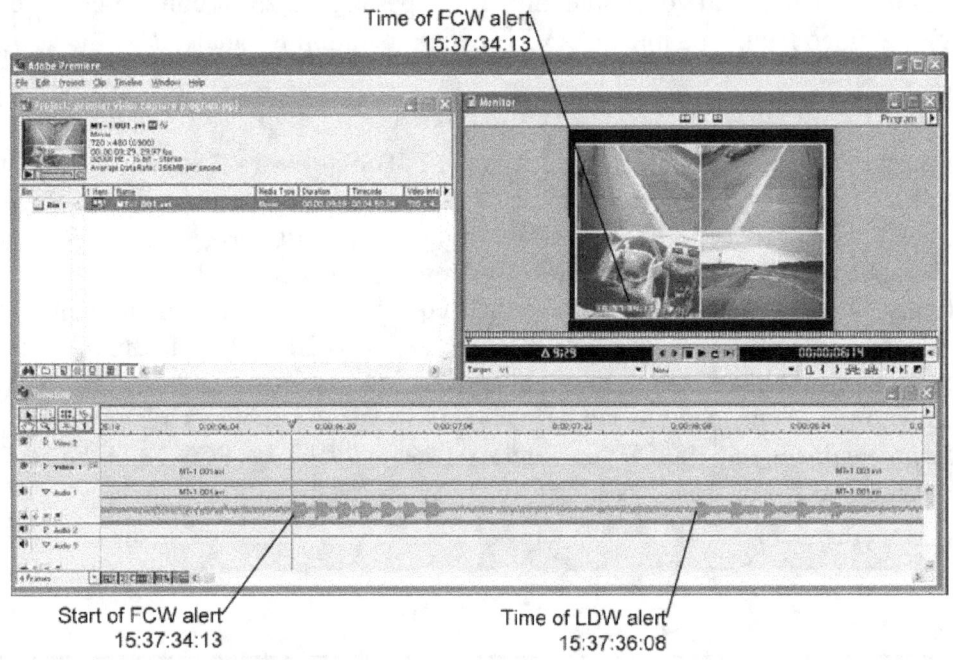

Figure 9. Locating time of warning using cab microphone audio.

7.2 Range and range-rate measurement (R_{FCW} and $Rdot_{SP}$)

The measurement system uses laser scanner data to determine forward range during a FCW warning (R_{FCW}) and range-rate ($Rdot_{SP}$). In order to obtain the correct range data, the laser-scanner data timestamp must correlate with the warning time. The laser scanner accepts GPS time updates, however, the resultant time stamp accuracy is not sufficient for data collected at the higher IVBSS test speeds. The effect of timing errors on range accuracy increases when the target is closing at higher speeds: at 20 m/s (about 45 mi/h) each 100 ms in timing error translates into a 2 meter range error. The process to overcome this timing error and to obtain an accurate range measurement uses the steps described below. Figure 10 aids the description and Appendix A explains the measurement uncertainty associated with measuring R_{FCW} in this manner.

1. Use the warning time obtained from the 30 Hz video (in this example, 15:37:34:13, see Figure 10 below) to locate the video frame closest to the warning time (15:37:34:12.). The original video time stamp captured at 30 Hz will typically not occur in the laser scanner synchronized video frame since the laser scanner software records the video frames at the lower resolution of 10 Hz.
2. Synchronize laser-scanner's video time stamp to agree with the GPS time stamp overlaid on the recorded video. The same time stamp correction also applies to the laser-scanner range data.
3. Locate and record the range to the POV in the laser-scan before the warning time. The best results occur using the closest range value in the cloud of range points returned from the POV.
4. Locate and record the range to the POV in the laser-scan after the warning time.
5. Use the warning time to compute a linear interpolation between POV range values. Report the result as R_{FCW}.
6. Use a second range reading to the target vehicle approximately 1 second earlier to calculate a differenced range-rate.

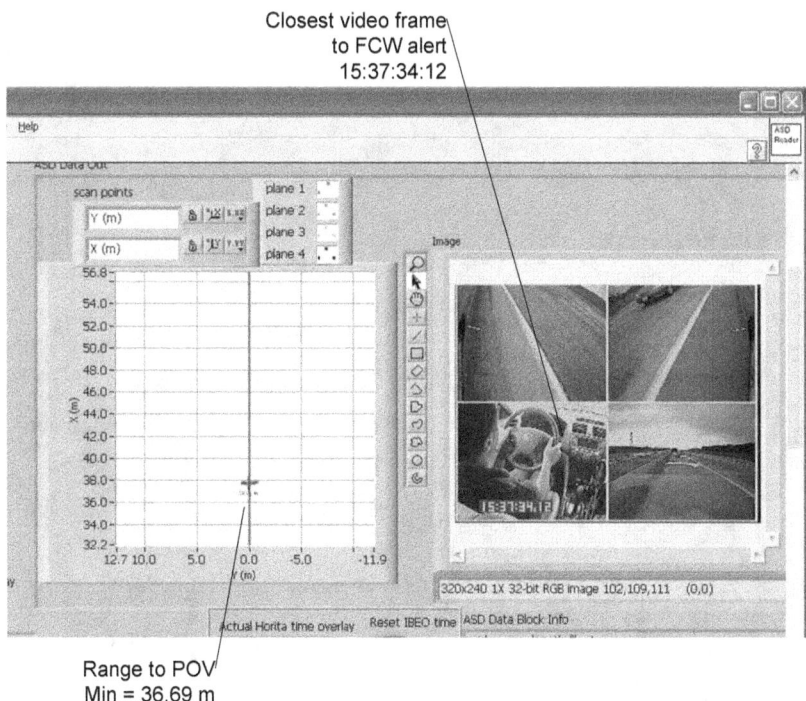

Figure 10. Measuring range at time of warning.

7.3 Lateral distance and lateral velocity (LatDist and LatVel)

Calibrated cameras provide data to compute lateral distance and lateral velocity to the lane boundary at the time of warning. No time correction is required since the measurements use the 30 Hz GPS time-stamped video rather than the 10Hz video associated with the range data. The lateral position uncertainty is estimated to be approximately ± 3 cm at 2 m from the wheel (see Appendix B) when the vehicle and the ground remain flat. In cases where the warning occurs with the vehicle outside the lane boundary when the side cameras cannot see the lane boundary, the forward view camera provides lateral position measurements, but with less accuracy and resolution.

To facilitate this lateral measurement process, NIST researchers developed software to step through an AVI file and to calculate the distance to any pixel in the calibrated region of the image (i.e., the region where calibration sticks initially laid). The process to measure lateral distance and lateral velocity consists of the following steps:
1. Locate the LDW warning frame (in this example, 15:37:36:08 in Figure 11.)
2. Select the pixel corresponding to the appropriate lane edge (left or right) and record the distance
3. Obtain a second distance to lane boundary measurement, just prior to the lateral warning time, to calculate a differenced lateral velocity

Figure 11. Measuring lateral distance to lane at time of warning.

8 Summary of Testing Activities

Phase I testing activities began in April 2007 and ended in March 2008. Testing activities conducted on each platform, shown in Figure 12, consisted of three distinct efforts: IMS integration/acceptance tests, dry run and witnessed verification tests, and Phase I extension tests. Table 2 highlights the extensiveness of the testing effort.

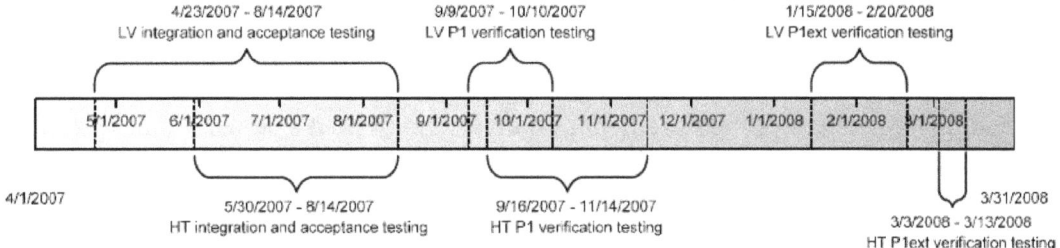

Figure 12. Light vehicle and heavy truck Phase I testing timeline.

Table 2 IVBSS Phase I Activities

#	Platform	Date	Location	Purpose
		IMS Preliminary Testing		
1	LV	April 23-24, 2007	Dana	Integration test
2	LV	May 29-30, 2007	TRC	Integration test
3	HT	May 30-31, 2007	Dana	Integration test
4	HT	August 13-14, 2007	Dana	Acceptance tests
5	LV	August 14-15, 2007	Dana	Acceptance tests
		Phase I Testing		
6	LV	September 9-12, 2007	TRC	Dry run track test
7	LV	September 12-13, 2007	Dana	Dry run track test
8	HT	September 16-21, 2007	Dana	Final track test
9	HT	September 24-25, 2007	Detroit	On-road test
10	LV	September 26-27, 2007	Dana	Final track test
11	LV	October 1-3, 2007	TRC	Final track test
12	HT	October 9, 2007	Marshall	Final track test (RD-3)
13	LV	October 10-11, 2007	Detroit	On-road test
14	HT	November 12, 2007	Detroit	On-road night retest
15	HT	November 13, 2007	Marshall	RD and LC/M retest
16	HT	November 14, 2007	Detroit	On-road day retest
		Phase I Extension Testing		
17	LV	January 15-17, 2008	TRC	Dry run track test
18	LV	February 4-6, 2008	TRC	Final track test
19	LV	February 19-20, 2008	Detroit	O-road test
20	HT	March 3-5, 2008	TRC	Dry run track test
21	HT	March 10-11, 2008	TRC	Final track test
22	HT	March 12-13, 2008	Detroit	On-road test

8.1 Measurement system acceptance testing

NIST researchers demonstrated the measurement system's accuracy and repeatability during a series of tests in August 2007. The primary concern was measuring R_{FCW} during tests conducted at high closing speeds. NIST developed a dynamic test to estimate the uncertainty of the IMS-measured R_{FCW}. The dynamic test, see Appendix A for details, analyzed the uncertainty in the laser-scanner range measurements, the time-stamp correction procedure, and the time-of-warning determination. The results indicated an uncertainty of approximately ± 1 m (95 % confidence) at ranges up to 60 m and at speeds up to 21 m/s (45 mi/h).

A second demonstration of measurement system accuracy used a series of lines painted at 1 meter intervals on the track surface as a ground-truth reference. A downward-looking camera captures the vehicle location relative to the reference marks and a microphone in the cab captures the audible warning. Figure 13 shows the reference marks on the ground along with the GPS time-stamped quad video. In this demonstration, a stopped vehicle sits at the 100 m mark (see lower right quadrant of Figure 13). Based on the ground marks, the warning took place between the 161 m and 162 m marks. This NIST measurement system reported the same result. A demonstration of the accuracy of range measurements for a moving POV used an additional downward-looking camera installed on the POV and a walkie-talkie to transmit the audible warning from the SV to the POV. With this setup, it is possible to measure the location of both vehicles with respect to ground marks at the time of warning. Again, the NIST measurement system agreed with the measurements obtained from the markings painted on the ground.

Figure 13. Upper right quadrant showing 1 m marks on track surface.

The project team also suggested a back-up procedure to double-check cases where the crash warning system's forward range measurements substantially disagreed with laser scanner measurements. To address this, NIST developed a simple calibration scheme of the forward camera to estimate the camera's focal length using the known POV width and the known distance from the camera to the POV. Figure 14 shows several images used to measure the range to the POV using the IMS forward camera. To calculate the range to the POV the user simply draws a box around the POV. Table 3 shows measured-range comparisons between the laser scanner and the forward camera. In most cases, the range measured by the laser scanner and the camera agreed. Run 5 shows one case where the camera vibrations made range measurement difficult; thus, it was not used.

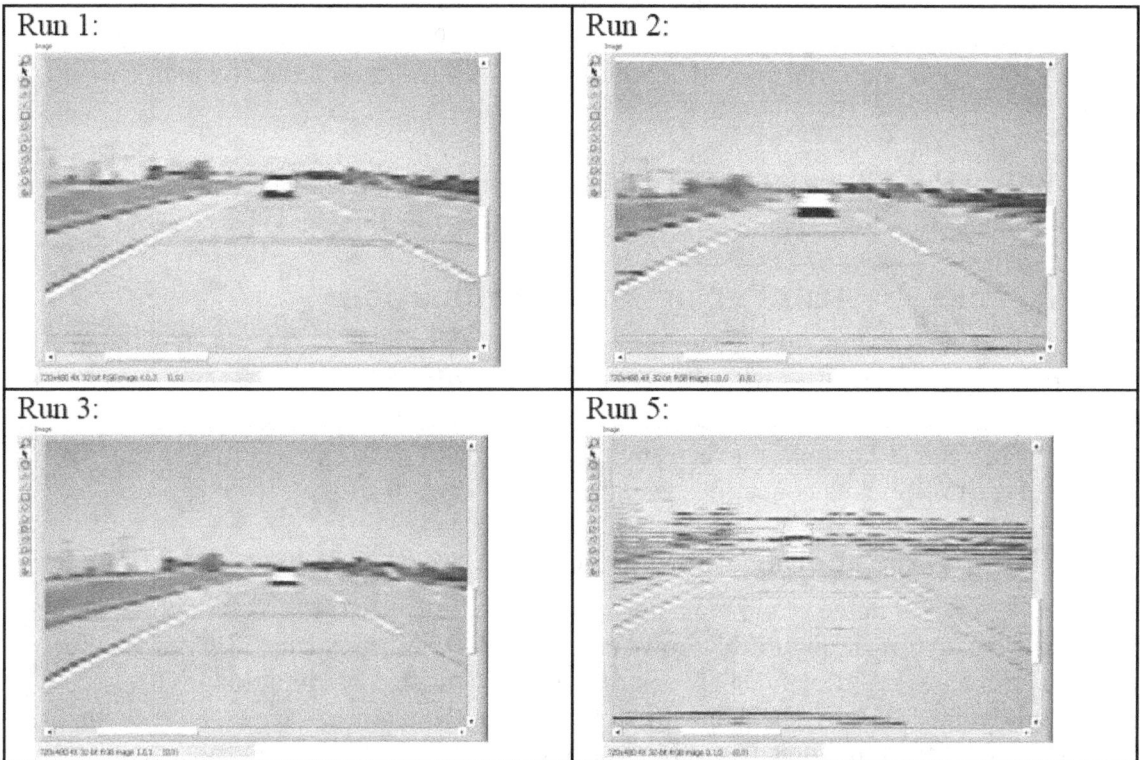

Figure 14. Example runs using the forward camera to measure range to the POV.

Table 3. Comparison between laser scanner range and forward-camera range.

Run #	LASER-SCANNER Range	Camera Range	Range Difference	Percent Difference
1	26.4	25.1	-1.4	-5 %
2	37.4	36.5	-0.9	-2 %
3	33.5	30.9	-2.6	-8 %
5	35.3	30.9	-4.4	-13 %

Tests demonstrating IMS accuracy ended in August 2007. At the conclusion of these tests, the IVBSS project team accepted the use of IMS measurements for judging warning system pass/fail.

8.2 IVBSS warning system testing

Phase I verification tests began in September and continued through November 2007. Testing included dry run practice tests, witnessed tests, and on-road testing. The U.S. DOT extended Phase I of the IVBSS program to allow team members to correct shortfalls in system performance identified during verification and on-road testing. Phase I extension testing, conducted between January and March 2008, verified that system changes resulted in improvement in system performance. At the conclusion of testing in March 2008, both platforms passed all closed-course verification tests and public road tests.

The IVBSS Phase I Interim Report and the Light Vehicle and Heavy Truck On-road Test Reports summarize the vehicle tests and discuss test results [9][10][11].

9 Warning System Performance Evaluation

This section describes analyzing and improving the performance of the crash warning system using data generated with the NIST IMS. The first example describes the construction of a model of warning range errors and using the model to correct range errors. The second example shows using the IMS data to uncover GPS time stamping errors in the IVBSS data collection system.

9.1 R_{FCW} error analysis and correction

Early tests showed range errors most noticeable with test scenarios involving high closing speeds. The top row in Figure 15 shows range errors from heavy truck dry run testing. Range errors are the difference between the forward collision warning range (R_{FCW}) captured by the on-board data acquisition system and the warning range (R_{IMS}) captured by the NIST IMS at the time of audible warning in the cab:

$$R_{err} = R_{FCW} - R_{IMS}$$

Based on an analysis of the errors the IVBSS team identified several errors, modified the data capture process and reduced warning system latency. In addition, NIST researchers developed the following error model from the range error in Figure 15a:

$$R_{err} = R_{FCW} - R_{IMS} = R_{off} + T_{delay} Rdot_{IMS} \qquad (1)$$

Where:
R_{FCW} = reported warning range (from on-board data acquisition system) (m)
R_{IMS} = laser scanner measured warning range (m)
R_{off} = a fixed offset between R_{FCW} and R_{IMS} (m)
T_{delay} = a timing delay between when a warning request is issued and the warning is heard in the cab(s)
$Rdot_{IMS}$ = IMS measured range-rate (m/s)

Using this model, the warning system can compensate for range error and warning delays using the following equation:

$$\hat{R}_{FCW} = R_{FCW} - R_{err} = R_{FCW} - \left(R_{off} + T_{delay} Rdot_{FCW}\right) \quad (2)$$

Where:

$\hat{R}_{FCW}$ = corrected value used by the warning system to trigger warning (m)
R_{FCW} = original uncorrected warning range (m)
R_{err} = see equation (1)
$Rdot_{FCW}$ = measured range-rate (m/s)

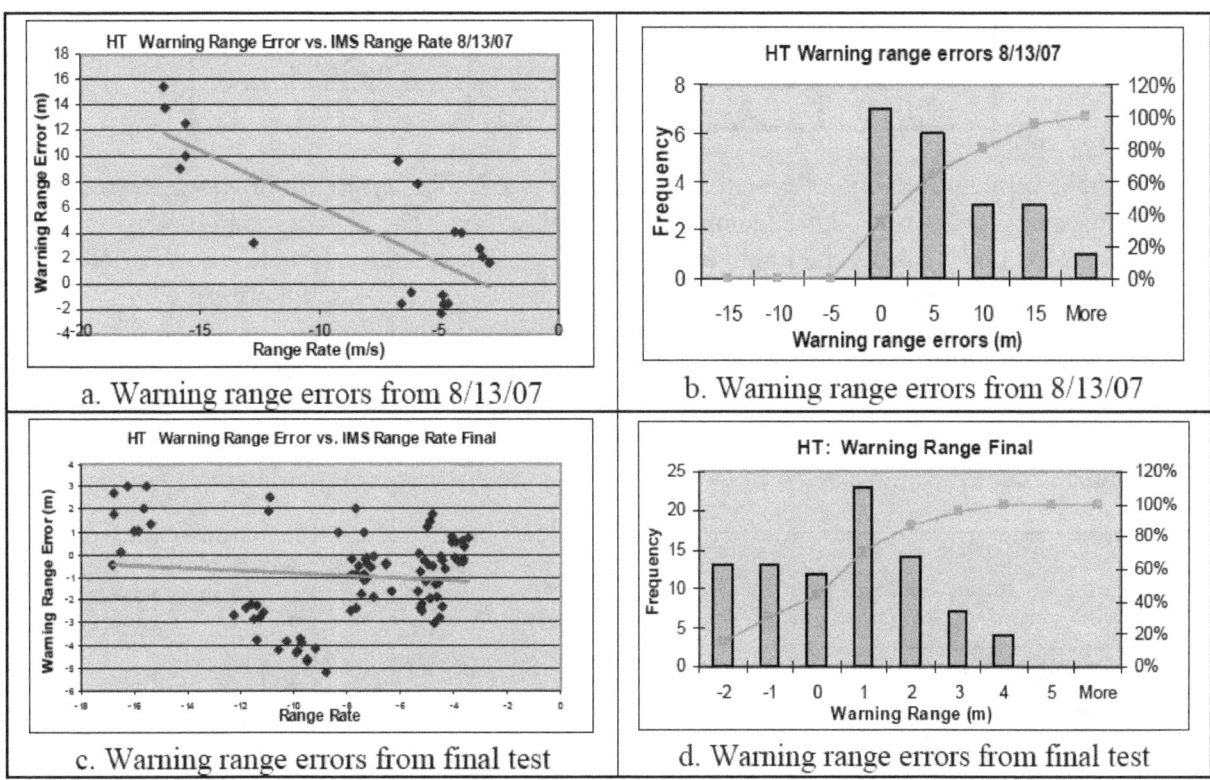

Figure 15. FCW warning range errors for heavy truck platform.

The second row in Figure 15 shows the warning range errors measured in the final heavy truck test series (R_{FCW}, used to compute the error, has an uncertainty of approximately ± 1 m (95 % confidence)). Table 4 summarizes the improvements, which include a 71 % reduction in warning range error and a 94 % reduction in the modeled warning-time latency errors (T_{delay}).

Table 4. Comparison between heavy truck dry run and final tests (average taken of |R$_{err}$|).

	Dry Run Range Error	Final Run Range Error	Percent Reduction Range Error
Average	5.33	1.55	-71 %
Standard Deviation	4.78	1.15	-76 %
Maximum	15.48	5.20	-66 %
Delay	-0.88	-0.05	-94 %
Offset	-2.71	-1.35	-50 %

The light vehicle team went through a similar process, implementing improvements to on-board data collection and reducing warning-delay time. The top row in Figure 16 shows R$_{FCW}$ errors during the light vehicle final tests. Again, there is a clear correlation between warning range error and range-rate. The light vehicle team did not have an opportunity to implement the error-model compensation technique. However, the bottom row in Figure 16 shows the possible results (obtained during post-processing) had the warning system used the technique. Table 5 summarizes the projected performance improvements realizable with NIST error-model compensation: a 60 % reduction in warning range error and a 96 % reduction in the modeled warning-time latency (T_{delay}).

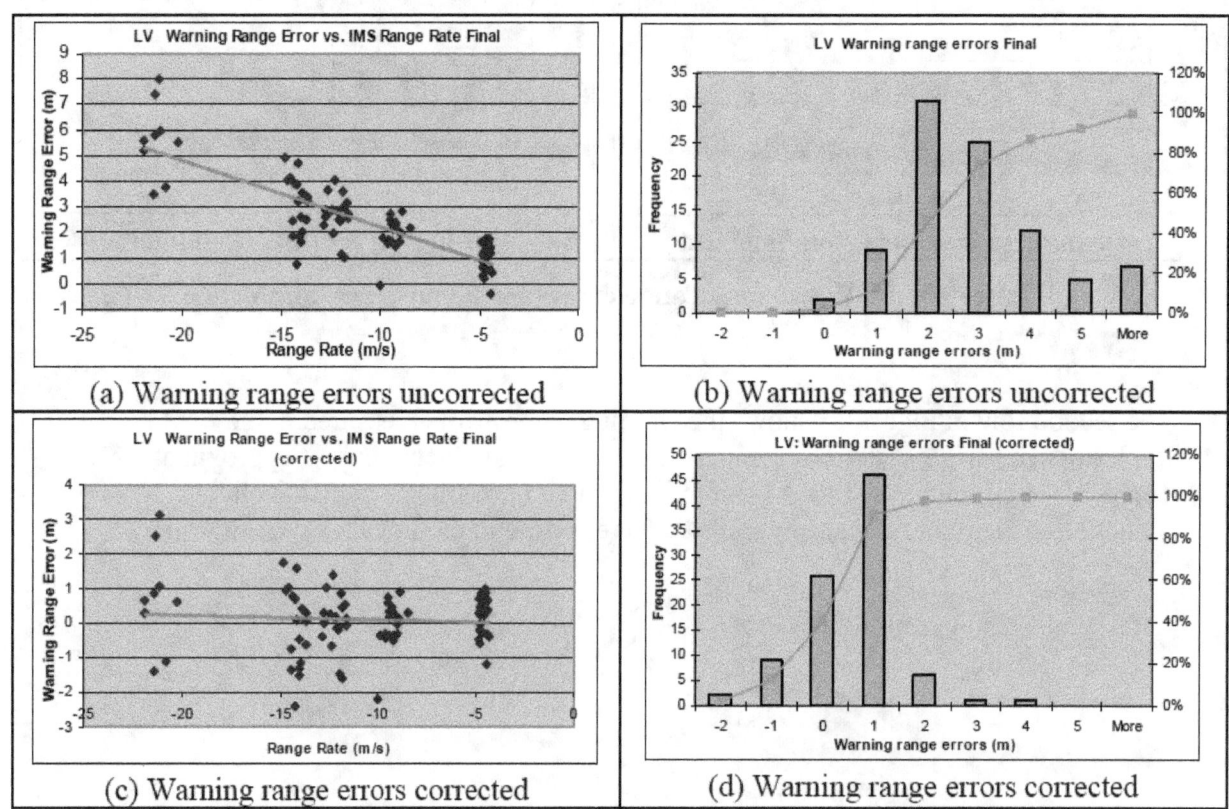

(a) Warning range errors uncorrected
(b) Warning range errors uncorrected
(c) Warning range errors corrected
(d) Warning range errors corrected

Figure 16. Comparison of light vehicle final test results uncorrected vs. corrected.

Table 5. Results if light vehicle final run were corrected (average taken of $|R_{err}|$).

	Uncorrected Range Error	Corrected Range Error	Percent Reduction Range Error
Average	2.238	0.673	-70 %
Standard Deviation	1.472	0.589	-60 %
Maximum	8.025	3.128	-61 %
Delay	-0.257	0.011	-96 %
Offset	-0.334	0.352	5 %

9.2 GPS timestamp error analysis

The NIST IMS and the test vehicle's on-board data acquisition system both use GPS to time stamp the time of warning. Figure 17 shows the errors (computed as the difference between the warning time reported by the on-board data acquisition system and that reported by the IMS GPS time stamp; a negative number indicates the test vehicle warning occurred earlier in GPS time). A plot over a 9 day period shows errors clustered in groups, with the error being consistent once the system was up and running. One possible source of this discrepancy may be an error in the initial synchronization with the GPS receivers. A detailed investigation of this error will take place prior to Phase II verification testing.

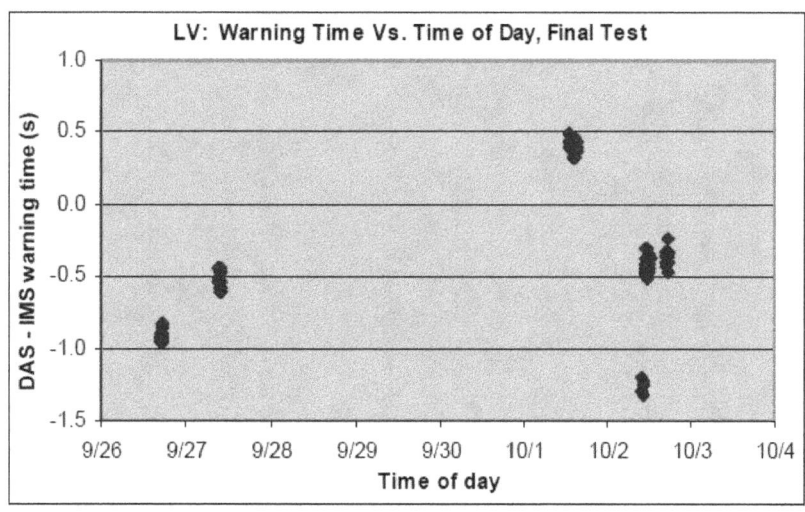

Figure 17. Error between on-board data acquisition system and NIST IMS reported warning time.

10 Summary

The independent measurement system developed by NIST is a real-time, vehicle-based system for measuring range and range-rate to objects surrounding a test vehicle and for measuring the vehicle's lateral distance and lateral velocity with respect to lane and road boundaries. IMS users can post-process and analyze the test data to achieve a high-degree of confidence in its accuracy and reliability. The system's forward-range uncertainty is approximately ± 1 m at distances up to 60 m and at target closing speeds of 20 m/s.

Independent measurements support the user in:
- Identifying errors in warning range
- Identifying warning system latencies
- Identifying errors in data time-stamping
- Modeling errors and applying the models to compensate for range and timing-delay errors

The system provides a wide range of measurement and data collection capabilities for both on-track and on-road testing without the need for instrumentation on other vehicles or of the roadway. A less-expensive approach such as using reference marks painted on the track surface or a calibrated forward-looking camera may suffice as an alternative or as a backup check for IMS malfunction.

11 References

[1] J. Ference, S. Szabo, W. Najm, "Objective Test Scenarios for Integrated Vehicle-Based Safety Systems", 20th International Technical Conference on Enhanced Safety of Vehicles (ESV), Lyon, France, June 18-21, 2007.
http://www.nhtsa.dot.gov/staticfiles/DOT/NHTSA/NRD/Multimedia/PDFs/Crash%20Avoidance/2007/Ference_Paper_07-0183.pdf

[2] J. Ference, S. Szabo, W. Najm, "Performance Evaluation of Integrated Vehicle-Based Safety Systems", PerMIS 06.
http://www.nhtsa.dot.gov/staticfiles/DOT/NHTSA/NRD/Multimedia/PDFs/Crash%20Avoidance/2006/PerMIS06_Ference-Szabo-Najm.pdf

[3] S. Szabo, R. Norcross, J. Falco, "Objective Test and Performance Measurement of Automotive Crash Warning Systems", International Society of Optical Engineering (SPIE) Defense & Security Symposium, Orlando, FL, April 2007.
http://www.itsa.org/itsa/files/pdf/Szabo%20SPIE%202007%2003-26-07.pdf

[4] "Integrated Vehicle-Based Safety System Light Vehicle Verification Test Plan", Prepared by Visteon Corporation for U.S. Department of Transportation, March 2008, UMTRI-2008-16. http://www.itsa.org/itsa/files/pdf/IntegratedVehicle-BasedSafetySystemsLightVehicleVerificationTestPlan.pdf

[5] "Integrated Vehicle-Based Safety Systems (IVBSS) Verification Test Plans for Heavy Trucks" Prepared by the University of Michigan Transportation Research Institute (UMTRI), Eaton Corporation, and Cognex Corporation for U.S. Department of Transportation. March 2008, UMTRI-2008-15.
http://www.itsa.org/itsa/files/pdf/IntegratedVehicle-BasedSafety%20SystemsHeavyTruckVerificationTestPlan.pdf

[6] S. Szabo, K. Murphy, M. Juberts, "The AUTONAV/DOT Project: Baseline Measurement System for Evaluation of Roadway Departure Warning Systems", DOT HS 808895, January 1999.
http://www.itsdocs.fhwa.dot.gov//JPODOCS/REPTS_TE//9623.pdf

[7] S. Szabo, R. Norcross, "Final Report: Objective Test Procedures for Road Departure Crash Warning Systems", DOT HS 810829, August 2007.
http://www.nhtsa.dot.gov/staticfiles/DOT/NHTSA/NRD/Multimedia/PDFs/Crash%20Avoidance/2007/ObjTestProc-RDCW-Final.pdf

[8] S. Szabo, B. Wilson, "Application of a Crash Prevention Boundary Metric to a Road Departure Warning System". Proceedings of the Performance Metrics for Intelligent Systems (PerMIS) Workshop, National Institute of Standards and Technology, Gaithersburg, MD, August 24-26, 2004.
http://www.isd.mel.nist.gov/documents/szabo/PerMIS04.pdf

[9] University of Michigan Transportation Research Institute (UMTRI), (May 2008). "Integrated Vehicle-Based Safety Systems (IVBSS) - Phase I Interim Report." Sponsored by National Highway Traffic Safety Administration, Washington D.C., DOT HS 810 952.
http://www.nhtsa.dot.gov/staticfiles/DOT/NHTSA/NRD/Multimedia/PDFs/Crash%20Avoidance/2008/810952Lo.pdf

[10] Harrington, R.J., Lam, A.H., Nodine, E.E., and Ference, J.J., Najm, W.G. (2008). Integrated Vehicle-Based Safety Systems Light Vehicle On-Road Verification Test Report (DOT HS 811 020). Washington, DC: U.S. Department of Transportation, National Highway Traffic Safety Administration.
http://www.nhtsa.dot.gov/staticfiles/DOT/NHTSA/NRD/Multimedia/PDFs/Crash%20Avoidance/2008/811020.pdf

[11] Harrington, R.J., Lam, A.H., Nodine, E.E., and Ference, J.J., Najm, W.G. (2008). Integrated Vehicle-Based Safety Systems Heavy Truck On-Road Verification Test Report (DOT HS 811 021). Washington, DC: U.S. Department of Transportation, National Highway Traffic Safety Administration.
http://www.nhtsa.dot.gov/staticfiles/DOT/NHTSA/NRD/Multimedia/PDFs/Crash%20Avoidance/2008/811021.pdf

[12] A. Lytle, S. Szabo, G., Cheok, K. Saidi, R. Norcross "Performance Evaluation of a 3D Imaging System for Vehicle Safety," in Proceedings of the Unmanned Systems Technology IX, Proceedings of SPIE, May 2007

[13] H. Scott, S. Szabo, "Evaluating the Performance of a Vehicle Pose Measurement System", Proceedings of the Performance Metrics for Intelligent Systems (PerMIS) Workshop, Gaithersburg, MD, August 13-15, 2002.
http://www.isd.mel.nist.gov/documents/scott/PerMIS_2002.pdf

Appendix A – Forward Warning Range Uncertainty (R_{FCW})

The process of measuring forward warning range at the time of warning (R_{FCW}) uses 30 Hz video and audio to capture the warning time and a commercially developed laser-scanner[3] to measure the range at the time of warning. The warning range is the ground truth for determining whether the prototype warning system passes or fails each test. Characterization of the laser scanner and the process to synchronize range readings with the time of warning revealed an R_{FCW} uncertainty of (at 95 % confidence):

$$U_{R_{FCW}} = 0.204\,\text{m} \pm 0.845\,\text{m} \tag{3}$$

Thus, at least 95 % of valid laser-scanner R_{FCW} measurements are within 1 m of the true warning ranges. This error bound is valid for the distance and speeds expected during the vehicle tests. The remainder of this section describes the process used to derive this error. Section A.1 discusses tests to measure laser scanner static range uncertainty. Section A.2 addresses dynamic range uncertainty (scanners are moving while measuring range), and warning time uncertainty (how precisely does the system detect the onset of a warning) is examined in Section A.3. Section A.4 summarizes overall R_{FCW} uncertainty.

A.1 Static range uncertainty

The IMS incorporates a pair of laser scanners to measure range to objects around the vehicle. To determine a performance baseline, the NIST team evaluated the laser scanner in a static environment using procedures developed for 3D measurement systems [12]. Measurements in a static environment isolate the scanner performance from the other sources of system uncertainty (e.g., mounting, target profile, and motion). The procedure evaluates the scanners separately against flat, white targets, and uses laser-surveying equipment to determine ground truth.

Figure 18. Static measurement setup.

---

[3] All vehicle tests used two commercially developed dual-head laser scanners mounted onto an instrument rack placed across the front of the test vehicle.

A.1.1 Procedure
Setup
The static evaluation uses a sensor mount, a target, and a survey device (see Figure 18). Normally the sensors mount horizontally on opposite sides of the test vehicle; in this case, sensors are mounted vertically to enable readings without lateral offsets. The target is a white board mounted on a vehicle. One side of the target has a coat of flat white paint and the other has strips of highly reflective tape. The two sides permit an evaluation of the effect of surface reflectivity on range readings. Survey equipment provides the ground truth for the measurements. A blanket isolates readings from the individual scanners in order to simplify processing. Figure 18 above shows the blanket on the upper scanner.

Data Collection
The static test procedure produces data at several distances, with alternative targets, and with varying angles between the target and the scanner. Figure 18 shows the test setup. The first step is to position the target at a desired distance and to survey the true position. With one scanner obstructed by the blanket, record two runs of data for two minutes each. Next, cover the other scanner with the blanket and record another set of two, two-minute runs. Finally, move the target to the next nominal distance and repeat the measurements. To complete the test, repeat the above steps for each side of the target (flat and taped) and with the scanners rotated by 90°.

Data Processing
Custom application software facilitates displaying the scan data and selecting data points that correspond to the target. Once selected, the software computes the center of the range values in the selected area and repeats the computation for the next 500 scans using the same selected area. The software reports the mean range value and standard deviation.

A.1.2 Results
Target Effect
Figure 19 is a box plot of the average range values separated by scanner and target type. The figure shows that there is a distinction between the two sensors and that surface reflectivity does not have an appreciable effect on range readings.

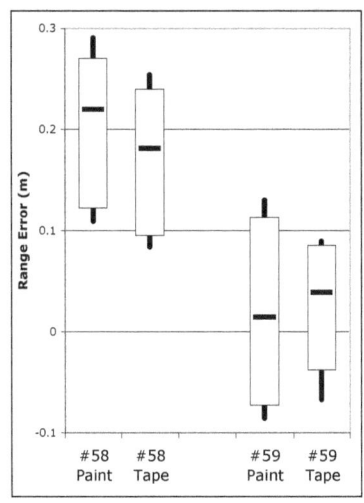

Figure 19. Effect of target reflectivity on Range Error.

Angle Effect

Figure 20 is a box plot of the average readings separated by sensor and relative angle. Other than an outlier with sensor #59, the data show the errors have less distribution at the rotated position but have minor effect on the expected value.

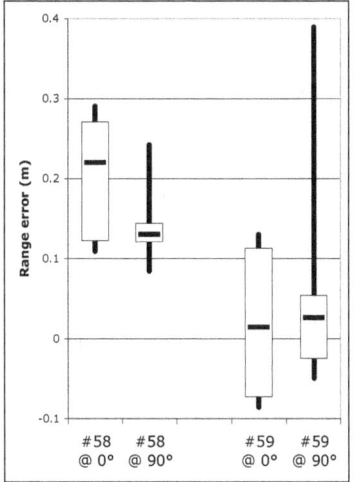

Figure 20. Effect of relative angle on Range Error.

Range Error

Figure 21 shows the range error as a function of range for the two sensors. As noted above, there is an offset between the two sensors but the trend lines are essentially parallel.

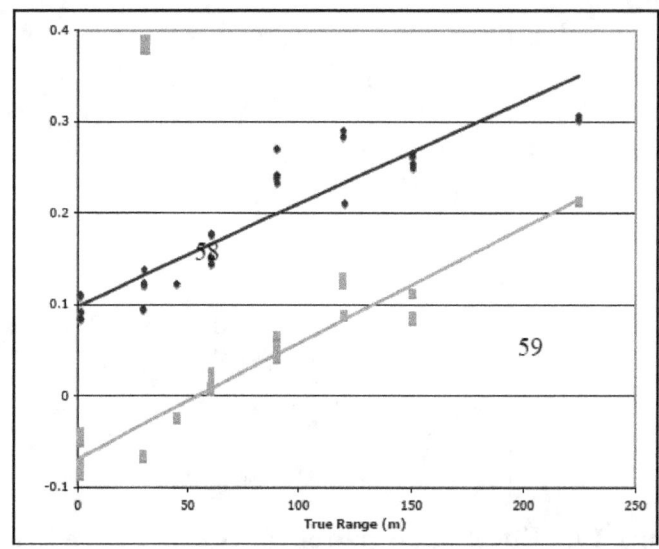

Figure 21. Range Error as a function of range.

For a Level of Confidence of 95 %, the measurement uncertainty is twice the standard deviation from the readings of each sensor:

$U_{58} = 0.150$ m

$U_{59} = 0.168$ m

A.2 Dynamic range uncertainty measurement

NIST researchers developed a dynamic test to determine range error and range uncertainty of the laser scanner when collecting measurements from a fast-moving vehicle. The test uses reflectors placed on the ground at surveyed distances approximately 20 m, 40 m and 60 m from a fixed target (see Figure 22). During the test, the vehicle travels over the reflectors at speeds of approximately 4 m/s, 9 m/s, 14 m/s, and 21 m/s (10 mi/h, 20 mi/h, 30 mi/h and 45 mi/h). An optical emitter/detector (1 ms uncertainty) mounted to the front bumper detects the reflector and generates a pulse. A GPS time-stamping unit latches the GPS time it receives the pulse (1 µs uncertainty) and generates an event time. Additional details describing use of this approach to characterize a GPS position measurement system appear in [13].

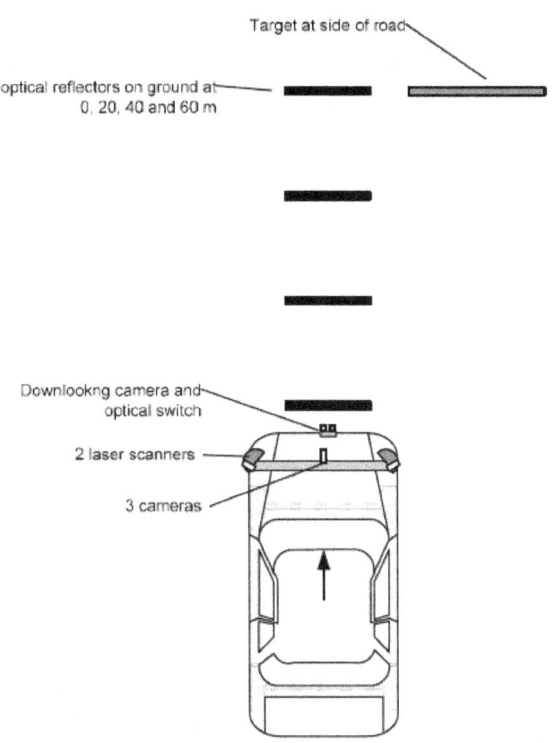

Figure 22. IMS Dynamic test layout.

The dynamic test produces a list of event times when the vehicle is a precise distance from a target, and a single file containing time-stamped laser-scanner data and video data. The procedure to correct GPS time-stamp errors and to measure range to the target at the event time consists of the following steps:

1. Use the GPS timestamp overlaid on the first video frame in the file to correct the laser-scanner data timestamp.
2. Locate the video frame closest to the event time (when the vehicle drives over the reflector).
3. Locate the range scan before the event time and record the minimum range value (R_B) and its timestamp (T_B).
4. Locate the next scan after the event time and record the minimum range value (R_A) and its timestamp (T_B).
5. Compute the interpolated range to the target at the event time using the following equation:

$$R_E = R_A + (T_E - T_A)\frac{(R_B - R_A)}{(T_B - T_A)} \qquad (4)$$

6. Calculate R_{diff} as the difference between R_E and the known distance to the target at the event time.

Figure 23 shows the range errors (R_{diff}) versus the true range from a number of runs conducted at various speeds. The data shows that range-related errors are insignificant (same conclusion from the static characterization).

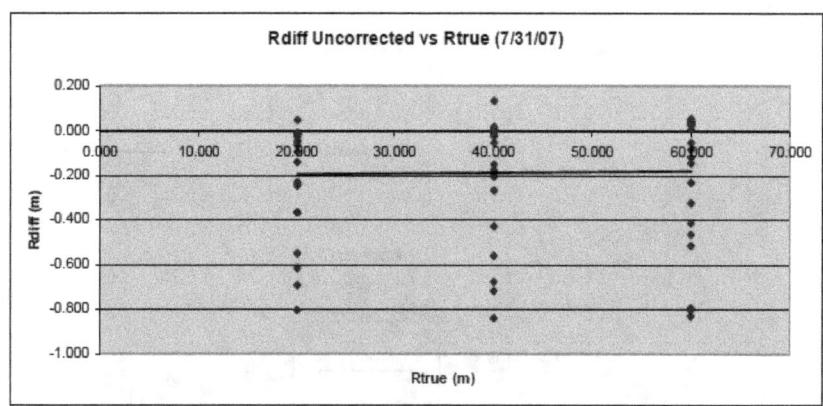

Figure 23. The range errors (Rdiff) vs. range.

Figure 24 shows the range errors (R_{diff}) versus the range-rate (R_{dot}). The data suggest that the process has minimal uncertainty due to timing delays or timing errors. A summary of the data appears in Table 6.

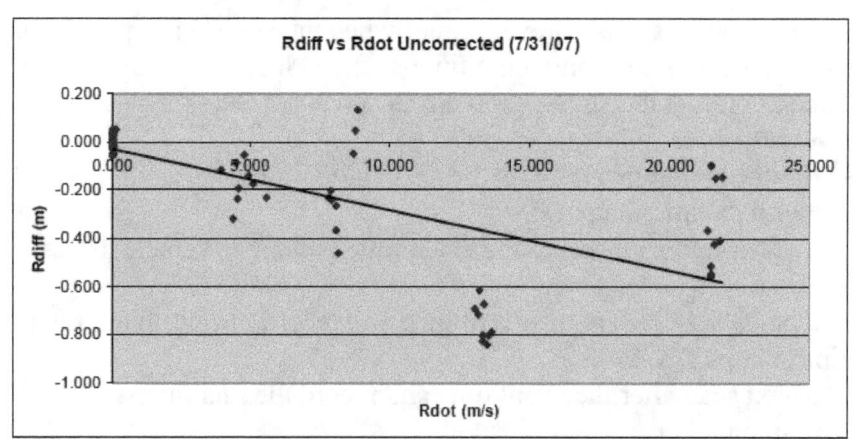

Figure 24. The range errors (R_{diff}) vs. range-rate (R_{dot}).

Table 6. Summary of dynamic range errors.

Statistics	ABS(R_{int}-R_{true}) m
Average (μ_{dyn})	0.204
Stdev (σ_{dyn})	0.257
Max	0.840

After correcting the time stamp and interpolating between range-scans, the dynamic uncertainty of the laser-scanner (95 % confidence) calculated from the measured data is:

$$U_{dyn} = \mu_{dyn} \pm 2\sigma_{dyn} = 0.204\,\text{m} \pm 0.514\,\text{m} \qquad (5)$$

A.3 Time of warning uncertainty

The warning system tests use a similar process as the dynamic test. The warning time (as opposed to the event time) is the GPS time on the video frame where the audible warning first occurs. The warning can occur at any point within a frame, so the worst-case uncertainty is ± 1 video frame (29.97 frames/s). The total uncertainty for determining the warning time is:

$$u_{wt} = \pm \frac{1\,\text{frame}}{29.97\,\text{frames/s}} = \pm 0.033\,\text{s} \qquad (6)$$

A.4 R_{FCW} combined uncertainty (dynamic and warning time)

The R_{FCW} warning range uncertainty includes the 30 Hz sample period used to determine the warning range time and the dynamic range errors of the laser-scanner. To combine these errors, warning time uncertainty is converted to a range uncertainty using the maximum longitudinal velocity the laser scanner encounters (in the stopped POV test, R_{dot_max} = 21 m/s (45 mi/h)). These uncertainties add as vectors (since they are orthogonally independent of each other) resulting in an overall uncertainty in R_{FCW} (95 % confidence) of:

$$U_{R_{FCW}} = \mu_{dyn} \pm \sqrt{(2\sigma_{dyn})^2 + (u_{wt} * R_{dot\_max})^2} \qquad (7)$$

$$U_{R_{FCW}} = 0.204 \pm \sqrt{(0.514)^2 + (0.033 * 21)^2}$$

$$U_{R_{FCW}} = 0.204\,\text{m} \pm 0.845\,\text{m}$$

Appendix B – Lateral Distance Uncertainty (LatDist)

Lateral distance measurements use calibrated side-looking cameras. Figure 25 shows calibration meter-sticks marked every 10 cm (2 m total) placed to the left and right side of the front wheels. To calibrate the video, NIST developed a software package that allows the user to select the pixels corresponding to each 10 cm segment. A table maintains the pixel to distance relationship. When the user picks a pixel between the red circles, the software automatically interpolates and computes the distance from the front wheel. The process only works for distance measurements on the road surface along the pixels marked by the red circles.

Figure 25. Calibration sticks with 10 cm marks adjacent to left and right front wheel.

Lateral distance measurements have the following uncertainties:
1. Camera optics – wider fields of view provide better coverage but produces radial distortions
2. Camera height and orientation – the perspective of a camera produce additional uncertainty that changes any time the camera moves
3. Image resolution – fewer pixels reduces the range resolution (only change with camera or with multiplexing scheme)
4. Operator pixel selection – selecting the appropriate pixel during calibration and during lane measurements affects the accuracy. The custom application software allows one to zoom in on the image, improving accuracy.

Table 7 shows an example of the resolution (in cm) for each pixel after camera calibration. The table illustrates that measurement resolution decreases as a function of distance from the vehicle wheel. Assuming a user can select the lane edge within 1 to 2 pixels, one can expect a LatDist uncertainty of ± 3 cm at 2 m (2 pixels * 1.31 cm/pixel).

Table 7. Resolution obtained from a calibration (cm/pixel) of side-looking camera

Distance from wheel (m)	Pixels/10 cm increment	Pixel resolution (cm/pixel)
0.1	26	0.38
0.2	26	0.38
0.3	23	0.43
0.4	24	0.42
0.5	23	0.43
0.6	23	0.43
0.7	19	0.52
0.8	20	0.50
0.9	15	0.65
1	17	0.58
1.1	15	0.65
1.2	11	0.88
1.3	13	0.76
1.4	10	0.96
1.5	9	1.08
1.6	9	1.05
1.7	8	1.12
1.8	7	1.41
1.9	8	1.24
2	7	1.31

www.ingramcontent.com/pod-product-compliance
Lightning Source LLC
Chambersburg PA
CBHW081807170526
45167CB00008B/3369